PETIT P●INT

A Candid Portrait on the Aberrations of Science

PETIT P●INT

A Candid Portrait on the Aberrations of Science

Pierre-Gilles de Gennes

World Scientific

DL Publishing

Published by

DL Publishing, an imprint of

World Scientific Publishing Co. Pte. Ltd.

5 Toh Tuck Link, Singapore 596224

USA office: 27 Warren Street, Suite 401-402, Hackensack, NJ 07601

UK office: 57 Shelton Street, Covent Garden, London WC2H 9HE

Library of Congress Cataloging-in-Publication Data
Gennes, Pierre-Gilles de
  Petit point : a candid portrait on the aberrations of science / Pierre-Gilles de Gennes.
    p. cm.
  ISBN 981-256-011-4 (pbk)
   1. Scientists--Attitudes. 2. Scientists--Humor. I. Title.

Q147.G4613 2004
502'.07--dc22   2004058368

British Library Cataloguing-in-Publication Data
A catalogue record for this book is available from the British Library.

Copyright © 2004 by World Scientific Publishing Co. Pte. Ltd.

All rights reserved. This book, or parts thereof, may not be reproduced in any form or by any means, electronic or mechanical, including photocopying, recording or any information storage and retrieval system now known or to be invented, without written permission from the Publisher.

For photocopying of material in this volume, please pay a copying fee through the Copyright Clearance Center, Inc., 222 Rosewood Drive, Danvers, MA 01923, USA. In this case permission to photocopy is not required from the publisher.

Printed in Singapore by World Scientific Printers (S) Pte Ltd

Contents

To the Minister of Research and Technology

Madam,

You look at Science from a top level and see it as a vast tapestry. I look at it from the bottom up and notice the details of every single stitch. Please allow me to present you with this collection of portraits. They are clumsy, often critical, but they are sincere: After a long life dedicated to science, I deeply love my heroes, even those I feel particularly bitter about. They live in a world of harsh competition and their weaknesses are all too natural. There is no need to dwell on the most obvious shortcomings: I will refrain from talking about

Mr. Grandpré who became so very formal the moment he joined the Academy, nor will I mention Mr. Buisson, who was so determined to play hero to student activists back in 1968.

I will, instead, paint more subtle cases where the negative and the positive are intimately intertwined — the genuine canvas of science.

Some people might cry foul and grumble that I dream of becoming another La Bruyère. I prefer to think of myself as an image hunter in search of a true moment in a scientist's life. I hope you will pardon my bluntness and that I may remain your very obedient and devoted servant.

astoc is a great czar of science. As a young man, he settled in an outlying province and immediately proceeded to infect novices with his contagious enthusiasm. He quickly embarked on a process of building. Within twenty years, he had given to his region a fully equipped laboratory that was well organized and respected on a national level. Some of his research topics turned out to be sterile, but all in all it did not matter much. Even if initially inspired by questionable motives, some of his actions ended up opening up brand new vistas. In more ways than one, Mastoc is a man of the past. He exercised absolute power over his fiefdom. Yet, that didn't prevent him from attracting excellent researchers who have gradually learned to assert their independence.

Young recruits have a great deal of difficulty accepting Mastoc's lordly style and they are quite vocal about their displeasure. Their criticism is somewhat unjustified because it ignores all the time he spent negotiating with regional authorities, fighting for his cause in Paris, planning the construction of new buildings, not to mention finding a job for every one of his students.

Post-war French science has often been rebuilt by the likes of Mastoc. But we should expect one last sacrifice from these pioneers: they ought to know enough to step aside gracefully as soon as their empire is solid enough to function on its own.

Véra came to us from Slavonia. She desperately wanted to work for the Lebel laboratory, but her expertise and degrees were less than stellar. The problem was easily solved: Véra had very attractive cheekbones and displayed the rather devastating charm her native country is renowned for. She proceeded to seduce Professor Lebel who promptly created an opening for her and found her a project. Her next challenge was to understand that project. Dutronc, a fundamental researcher, eagerly provided help in exchange for

Véra's favors. Unfortunately, scientific life is demanding: Véra's success hinged on building (or at least improving) a delicate measuring instrument. Bolet, a wizard in the machine shop, offered himself as the next sacrificial lamb.

The edifice of science is complex and there are many ways to make one's way across it. But Véra's journey was cut short. Irène, Professor Lebel's previous lover, made sure that order was restored.

Lanterne's primary job is to milk the cows. Not the cows grazing in the fields, mind you. Rather, I'm talking about the fat cows whose milk nurtures European science in the form of grants. The wise men in Brussels put out word that a particular program is deemed important and will be funded. Lanterne quickly hears of the opportunity. In just a few months, he lines up a team of researchers from established laboratories in Germany, England, and France. For good measure, he adds a few more groups of perhaps lesser stature but strategically based in Southern Europe. He puts together a project that is sure to win approval and to be allocated a sizeable amount of money. It does not matter that Lanterne does not have a precise vision of what the funding agency really wants (in terms of fundamental science or industrial payoff). He is, however, an undisputed master when it comes to reading annoucements issued by Brussels, creating organizational charts, and pumping for grant money.

Never mind that the funding he secures hardly ever addresses the real needs of the research team. More often than not, it goes toward risky endeavors such as directing a Europe-wide thesis involving contributors scattered all the way from Sicily to Scotland. The project may be shaky, but Brussels is satisfied anyway and the cows are happily milked.

Leduc used to be a very charming and enthusiastic man who introduced a new scientific field in France. Alas! His charm earned him a regular spot before television cameras. He was constantly being interviewed on a wide range of topics: not only those he knew but also those he didn't. Under the intoxicating heat of the spotlights, Leduc strayed towards unfamiliar territory and committed serious blunders.

Conclusion: It behooves us each of us to stay put in our own holding pen, like a retired thoroughbred put out to stud. And should an attractive filly from the media come prancing by the fence, by all means try to stay calm.

mmy is an American. She is a researcher and proud to be one. She worked day and night to complete her thesis. In her own country, women scientists are very much in the minority despite being in great demand. To prove herself, Emmy is determined to set up and run a scientific group. She gives her all toward that goal. Her husband lives a thousand kilometers away; she gets to see him only once a week. Needless to say, she has no children; her first priority is to succeed in her career and to do so she has to work around the clock.

Her life is a nightmare and her results often suffer from her exhaustion.

This state of affairs would change if there were more American women in science. But female students are well aware of the situation I'm talking about and, not surprisingly, they tend to choose other careers. The inevitable result is a system that is paralyzed.

reton had an uncanny ability to grasp things quickly and to think clearly. It was always a real pleasure to discuss with him a problem still in its embryonic stage. But Breton was born a little too early. He was trained as a physicist in the immediate aftermath of World War II — a time when our country desperately lacked experts.

Breton was like a gifted cello player who had never taken the time to practice the scales. His lack of finger dexterity did not prevent him from setting up — all by himself — a small but remarkably creative team. He, on other hand, never truly blossomed. There are many scientists of the same generation who, like Breton, left us unfinished symphonies. Meanwhile, scientists like myself, a little younger, naïvely pat ourselves on the back for having successfully composed three decent bars in more propitious times.

S mirnoff was trained in the rigid framework of a Soviet university. He had developed a real strength in mathematical analysis as well as a certain distance from the real world. All his papers begin with: *Izviestno,* "*It is well known that...*". They are virtually impossible to digest. Smirnoff would never stoop so low as to explain his goal, his technique, or what he was not completely certain about. Yet, one must admit that he can from time to time come up with some very good ideas.

To get the Smirnoff tree to really start bearing fruits, one would have to uproot it and transplant it to the United States. There, after a suitable incubation period, it would open itself to the world, regain its senses, strive to choose problems more judiciously, and make the effort to explain them better.

Be careful, though: it is best to transplant a single tree rather than a whole forest. An overly in-bred Russian colony feeding on its nostalgia for the "old country" can quickly turn into a stifling environment — at least as far as science is concerned (I'm not thinking of Nina Berberova or Vladimir Nabokov here).

Pluvieux was once an inspiring innovator who injected renewed vigor into an aging field of science. Sadly, he squeezed his way into a government agency where he spent the greater part of his mature years. He tried to build something in the face of constantly changing ministerial directives, of bureaucratic inertia, and of the recriminations of many end users.

The ordeal just about drained him. We would all have been better off if Pluvieux had never left research and had allowed himself the time to pause and think; if he had returned to academia and chosen a totally new scientific domain. To be sure, going back to the starting point is painful. Science is not unlike a game of snakes and ladders.

éziers is one of the great theorists of our time. He developed wonderful and far-reaching concepts. Unfortunately, Béziers decided at some point that he was destined to be a prophet. He would deal with human problems with the same cold rigor and insulation from everyday life that had served him so well in the abstract sciences. Rigidity and blindness often caused him to shoot from the hip and make snap decisions, even in public. Thank heavens Béziers was born in the 20th century and not at the time of the Inquisition.

uba was once the paramount magician among experimentalists. Cooped up for many years in a humble wooden cabin, he built exquisite machines from which he extracted important findings. Unfortunately, fame came his way and Kuba ended up landing in the comfort of a wealthy university campus, dealing with major contracts and an army of students. His role was reduced to describing what his troops were busy doing. He is a brilliant speaker and has a marvelous sense of humor. Blessed with such gifts, he can afford to coast without having to do any real work. Hopefully, lightning will strike him one of these days and force him to recover his eyesight.

Vladimir dedicated his youth to an esoteric science. He even created a whole new school of thought. Unfortunately, he fell into two traps: the first was success (a considerable success at that) and the second was the temptation to predict the future (a normal consequence of success). When Vladimir showed up on a university campus, crowds gathered around him. He would then deliver some obscure message combining formality and philosophy. At one time, for instance, Vladimir had come to believe that he had discovered the profound nature of life. His pronouncements had no basis in fact but, sadly, they lured many young people towards nebulous programs, from which it took them a long time to extricate themselves.

It is dangerous to mistake a researcher for a prophet. And we French people are particularly vulnerable on that front. We are enamored of discourse and factions. Many of our predecessors came out in favor of Descartes' risky mechanical models instead of Newton's accurate science. And more recently, some of our intellectuals passionately defended

Jacques Benveniste in the face of all common sense.[*]

[*] Dr. Jacques Benveniste was a French biochemist who claimed that certain immuno-tests worked at homeopathic dilution. Later, he also maintained that pure water has a special form of memory. He was not supported by the scientific community, but he was defended loudly by French intellectuals who insisted on "the right to heresy".

glaé walks into the conference room wearing a large hat that fully covers her eyes. She attracts everyone's attention. She wears the flashy headgear with supreme elegance.

The first impression most of us get is that she is a frivolous woman out to seduce. Watch out, though; that would be seriously misjudging Aglaé. In reality, she is a veritable lioness relentlessly focused on the tough task of organizing scientific research.

S ubtil emigrated long ago from Central Europe to the New World. He dedicated himself to a quest for the Holy Grail. His self-appointed mission was to invent new materials with exceptional properties. And with his incredible flair he did indeed make a number of tantalizing discoveries. They came close to the elusive Grail, but they were never quite the real thing.

His life was like a storm. In the tradition of knights of yore, he seduced many women along the way to add excitement to his quest, without much concern for the damage he caused in his wake. His sense of humor and his good knowledge of the old world brought him success. Perhaps the man was at once perturbed and driven by the unfinished aspect of his great work.

I have always been very fond of Subtil and the end of his story fills me with great sadness: the real Grail ended up being discovered not long after his death by other knights who had chosen to follow completely different trails.

C hazot reminds me of a butterfly flitting from flower to flower across the field of science. To be fair, he is a learned butterfly. He is always up on the latest fashionable theory and he understands it well enough. The opening of a Chazot paper echoes like fireworks with all manner of firecrackers exploding all at once. For instance he boldly links earthquakes and biological mutations. Chazot then proceeds to inject the theory everyone is buzzing about into the mix, expanding it to encompass the entire universe.

What is the long-term legacy of such constructs? Some good things and some not so good. Beginning scientists are apt to be dazzled by the scope of the questions posed and find it a source of inspiration. In the end, Chazot's real vocation is perhaps to give talks to high school students.

nchor has worked forever for a major research agency where he carried out some useful experiments. But this was not enough to fulfill him. He got himself appointed to one of the myriad French committees whose purpose is to monitor the research of others. There, he judges and renders summary verdicts. Most of the time, the projects under review extend well beyond his own knowledge. In no way does that stop him. With pompous pronouncements, he anoints himself the ultimate arbiter of elegance. Other less confident committee members do not dare protest.

This is how the future of a promising research area finds itself at the mercy of some obscure critic. When it comes down to it, research programs and theater plays are judged in much the same way.

Croesus used to be an imaginative researcher and a brilliant teacher. Why did he become so obsessed with money in his later years? The system he set up relies mainly on students from the Third World. Each of them is assigned an applied research project funded by an industrial contract that supports the facilities, provides a small stipend for the student and a rather generous salary for the principal investigator. And some of us are naïve enough to believe that slavery has been eradicated in the Western world!

C aesar started out as a brilliant French high school student, equally at ease in literature and the sciences. Later on, he became fascinated in some unusual geometrical objects invented in the early 20th century by mathematicians (also studied by several great physicists).

Caesar's claim to fame is to have understood their importance in our everyday life. Nature provides us with many examples of such objects, from the infinitesimally small to the infinitely large. Caesar's books made it a lot easier for us to understand them. Not to be trifled with in today's culture, Caesar came up with a clever name to describe these objects — a name that stuck and has been adopted all over the world.

Caesar has clearly enriched science, even though some physicists downplay his personal contribution. But success also has a down side: Caesar has become uncompromising and difficult. He believes he is the keeper of the Tablets of the Law. Whenever an article in his field fails to cite him prominently, he criticizes it viciously. Embittered prophets are as common in modern science as they were in the Old Testament.

Guru has had a long career in industrial research. Start talking about a difficult technical problem, and he will invariably profess to know all about it.

Sometimes a young researcher comes to Guru to suggest a radical change in a product formulation or in its manufacturing process. The inevitable answer is: "That will never work. We tried it already in 1967 and the reactor became hopelessly clogged."

Men of that type often kill innovating ideas. But let's be fair: Guru is one of the very few individuals intimately familiar with his company's know-how. He is as necessary and abusive as a sorcerer in an African tribe. Who among us could write about the ethnography of scientific gurus?

Dourakine was a great scientist. He used to reign supreme over an important branch of physics. We owe him much. Just the same, Dourakine was prone to losing his bearing: He wrote entire books based on a false premise. His hope was that, once an idea of his was enshrined in a book, it would become accepted once and for all.

Dourakine was a man to be feared. One day, he became miffed at the perceived impertinence of Nafion, a young foreigner. He promptly wrote a letter to Nafion's boss demanding the young man's dismissal. Thankfully, Nafion was able to defend himself, but it was a hard blow.

Still, we should pay tribute to Dourakine for surviving Stalin's years and contributing to the progress of science, even though his methods were often inspired by the ruler of his native country.

S aplir's main purpose in life is to organize con- ferences. For the last twenty years, he has been active in a particular field (without ever having personally produced any notably new results). Saplir gathers a committee, comes up with a title, chooses a date and place to hold the next meeting, and sends out *missi dominici* to get financial support from various industrial firms. The conference is usually held in a luxury hotel near an attractive beach. Sometimes, the conference even disseminates interesting new developments.

Whether good or not, the event always generates a book of proceedings with Saplir's name conspicuously plastered on the cover. And with yet another symposium under his belt, Saplir is ready to do it all over again the following year.

This system has some unfortunate consequences. I'm thinking, for example, about Naif, a young researcher who is working his heart out in a remote province on an aging topic for which Saplir is the eternal flag bearer. Naif is invited to the conference. Dazzled by the splendor of the accommodations and the many different nationalities of the attendees, he sees his self-confidence boosted to new heights. He returns home convinced that his topic is really hot. Only to squander the next ten years of his life toiling in a dead end.

anfred is one of the most original researchers of his time. He has solved a number of problems. He has invented a highly practical device now used all around the world. Manfred is also credited with having opened up another field that seemed initially quite forbidding and is now flourishing. But what a weird spirit he is! He can spend unlimited energy and waste years of his life on a trivial controversy. What's more, he holds long-standing grudges, which doesn't help his image.

His convictions can be quite bizarre as well. This champion of the exact sciences remains convinced to this day that, using nothing but will power, an individual has the ability to bend a spoon at a distance. Contrary to popular belief, some great scientists rely more on passion than on logic.

obert is curious about everything. As a young theorist, he spent much time working on arcane calculations. Eventually, his primary interest switched to pedagogical issues. He committed himself to finding for any phenomenon, no matter how complex, one simple explanation that anyone, even a layman, could understand. This kind of approach is undoubtedly useful, but it has progressively caused Robert's mind to undergo a mutation of sorts. When faced with a new problem, Robert disdains a rigorous approach. Instead, he favors educated guesses and qualitative estimates. He is turning more and more into something of an angler. Once in a great while, he still manages to reel in a nice fish. Most of the time, however, his catch is meager. But life goes on and Robert is quite happy standing on the banks of his own river.

olymorph directs a large French research agency. He puts all his energy into it. Over the years, the bureaucrats around him have inexorably swallowed him and their doctrine has become his own. To take a concrete example, researchers between the ages of thirty-five and forty-five face endless competition and experience enormous pressure. Coexistence does not come easy to them. Their preferred modus operandi is to head small, independent groups. Yet, Polymorph does not see it that way; he orders a reorganization which merges several groups into one single, large laboratory. It doesn't take long for the consequences to surface. Predictably, group A starts feuding with group B. The only way to avoid disaster would be to separate them. But Polymorph and his associates just don't care. They continue to insist on large organizational structures and to force marriages against nature. Why? Simply because it is much easier for central agencies to manage larger entities.

The ship of science is very heavy and no iceberg will make it alter its course.

Révizor was educated in what we call in France a "Grande Ecole"[†]. He produced a theoretical thesis that fortunately no one remembers today. His true vocation took shape not long thereafter. He went on to manage a series of major state agencies. The fields he has had to deal with have nothing whatsoever to do with what he was taught at

---

[†]"Grande Ecole" in France is a highly selective institution of higher learning in which students are accepted upon passing a rigorous entrance examination following an intensive two-year preparatory program.

the university. But he proves very adept at negotiating important agreements with foreign countries, encouraging innovation, fostering interdisciplinary programs, and other politically correct paradigms of our times.

Révizor is not as fierce as his name might suggest. He is a man of consensus. Having essentially given up on trying to understand things, he is a prime candidate for management.

Feston was in charge of a graduate program. His field of expertise had become conventional and a bit dusty.

With unbelievable energy, Feston expanded the scope of his curriculum three-fold within just a few years. He selected a staff of young and lively lecturers. The degree awarded by his program has since degenerated into a "degree in everything". It has enjoyed considerable success, particularly among many students from Latin countries, who prefer to remain generalists rather than focus on a specialized area.

Ask any graduate of this system whether he is familiar with different branches of physics, such as field theory, polymer science or biophysics. He will confidently answer yes on all counts. In reality, his knowledge about these subjects is barely skin deep.

How should one assess Feston's system? If you accuse him of superficiality, he will protest that he is simply promoting interdisciplinary research. While I do believe that Feston is an energetic and honest man, I'm afraid that his approach is bound to produce rather shallow researchers.

hilostrate is a professor at a major French university. As the years went by, he lost his interest in science and fell victim to the dreaded disease of Committee Meetings. The issues debated (in a Florentine style) in such meetings range from promotions to financial grants, which most of the time are aimed at important laboratories that are well established and belong in the mainstream. These meetings also deal with the planning of teaching loads. In keeping with tradition, freshly graduated high school students — the most challenging group — are routinely assigned to junior teachers, whereas the (rare) students who have climbed the upper rungs of the ladder are the only ones who get to face the senior professors.

Philostrate navigates his way across this strange world. He is at once vague about his goals and adamant about his rights. The young students who work in his archaic laboratory are soon appointed lecturers. Philostrate prudently avoids hiring outside researchers who might be too outspoken for his taste. He is even suspicious of neighboring universities. He makes sure (with the full support of the Science Ministry) that his own campus retains control of their post-graduate programs.

Knowledge suffers in the process, but peace is maintained.

Why has Philostrate's world evolved that way whereas many other Western countries boast highly creative universities? It would be unfair to place all the blame on Philostrate himself and his lobby. The real culprit is a government policy that guarantees any high school student admission to the university closest to his or her home, without any prior screening. After years of teaching poorly motivated students, Philostrate has simply lost his faith.

lise has struggled all her life. Born in an isolated area of the Balkans, she subsequently divided her time between Israel and France. She worked extra hard to defend her thesis. The field she had chosen was not trendy. Nevertheless she was able to develop highly original methods and ideas.

Fate wasn't kind to her. Her laboratory, buried deep in the bowels of a building, had all the charms of a tomb. More importantly, Élise was not part of the exclusive circle of the "Grandes Ecoles". She could only recruit students of limited ambition. Against all odds, she managed to establish strong contacts all over the world and she rekindled the flame in her students by setting up mutually beneficial collaborations.

It wasn't until much later that the rest of us science pedestrians came to appreciate the trail she had blazed. Élise never got the public recognition she rightfully deserved. At the end of the day, does it really matter? She is by now a grandmother and perfectly happy.

S piros holds court in Brussels, where he heads a branch of European science through a complex administration. To reach such a position, he had to prove his determination and his talent as a speaker. Scientific ability, on the other hand, was conspicuously absent from the list of prerequisites. Spiros bases all his decisions on reports that are themselves based on other reports or on a few visits to major industrial conglomerates. He sets up very costly and large programs of applied research controlled by these conglomerates.

Such programs are highly unlikely to provide the world with tomorrow's food or the high-speed trains we all need. In truth, these are the kind of programs that industry is generally suspicious of but decides to initiate anyway — as long as the funding is generous.

The management of science in Europe does not measure up to scientific research in Europe.

Akbar is a Brahman. He was trained in England in the applied sciences but he quickly found himself drawn towards the most difficult kind of theoretical work. His output is considerable and his modesty excessive. We hardly ever hear about him. On the American campus where he has decided to live, he is known among his peers mostly for his great artistic sense and his elegant style of trout fishing.

Akbar is an inspiration for us all; for his exquisite science, for his unassuming demeanor, and for the admirable way he has blended three different cultures.